Discover Health
Student Workbook

AGS®
American Guidance Service, Inc.
Circle Pines, Minnesota 55014-1796
800-328-2560

© 2000 AGS® American Guidance Service, Inc., Circle Pines, MN, 55014-1796.
All rights reserved, including translation. No part of this publication may be
reproduced or transmitted in any form or by any means without written permission
from the publisher.

Printed in the United States of America

ISBN 0-7854-1846-6

Product Number 91053

A 0 9 8 7 6 5 4

Table of Contents

Workbook Activity	1	Cells, Tissues, and Organs
Workbook Activity	2	The Body's Protective Covering
Workbook Activity	3	The Skeletal and Muscular Systems
Workbook Activity	4	The Digestive and Excretory Systems
Workbook Activity	5	The Respiratory and Circulatory Systems
Workbook Activity	6	The Nervous System
Workbook Activity	7	The Endocrine System
Workbook Activity	8	The Reproductive System
Workbook Activity	9	Hygiene
Workbook Activity	10	Fitness
Workbook Activity	11	The Family Life Cycle
Workbook Activity	12	Dealing With Family Problems
Workbook Activity	13	Emotions and Their Causes
Workbook Activity	14	Social Emotions
Workbook Activity	15	Emotions and Behavior
Workbook Activity	16	Managing Frustration
Workbook Activity	17	Managing Stress and Anxiety
Workbook Activity	18	Responding to Peer Pressure
Workbook Activity	19	Eating Disorders
Workbook Activity	20	Being a Friend to Yourself
Workbook Activity	21	Making Friends
Workbook Activity	22	Healthy Relationships
Workbook Activity	23	Food for Energy
Workbook Activity	24	Carbohydrates and Protein
Workbook Activity	25	Fats and Cholesterol
Workbook Activity	26	Vitamins, Minerals, and Water
Workbook Activity	27	Dietary Guidelines
Workbook Activity	28	Food Choices and Health
Workbook Activity	29	Influences on Food Choices
Workbook Activity	30	Reading Food Labels
Workbook Activity	31	Making Foods Safe
Workbook Activity	32	Prescription and Over-the-Counter Medicines
Workbook Activity	33	The Effect of Medicines and Drugs on the Body
Workbook Activity	34	Tobacco
Workbook Activity	35	Alcohol

Workbook Activity	36	Narcotics, Depressants, Stimulants, and Hallucinogens
Workbook Activity	37	Other Dangerous Drugs
Workbook Activity	38	The Problems of Drug Dependence
Workbook Activity	39	Solutions to Drug Dependence
Workbook Activity	40	Causes of Disease
Workbook Activity	41	The Body's Protection From Disease
Workbook Activity	42	AIDS
Workbook Activity	43	Sexually Transmitted Diseases
Workbook Activity	44	Cardiovascular Diseases and Disorders
Workbook Activity	45	Cancer
Workbook Activity	46	Asthma
Workbook Activity	47	Diabetes
Workbook Activity	48	Arthritis
Workbook Activity	49	Epilepsy
Workbook Activity	50	Promoting Safety
Workbook Activity	51	Reducing Risks of Fire
Workbook Activity	52	Safety for Teens
Workbook Activity	53	Emergency Equipment
Workbook Activity	54	Safety During Natural Disasters
Workbook Activity	55	What to Do First
Workbook Activity	56	Caring for Common Injuries
Workbook Activity	57	First Aid for Bleeding, Shock, and Choking
Workbook Activity	58	First Aid for Heart Attacks and Poisoning
Workbook Activity	59	Defining Violence
Workbook Activity	60	Causes of Violence
Workbook Activity	61	Preventing Violence
Workbook Activity	62	Health Care Information
Workbook Activity	63	Seeking Health Care
Workbook Activity	64	Paying for Health Care
Workbook Activity	65	Being a Wise Consumer
Workbook Activity	66	Evaluating Advertisements
Workbook Activity	67	Consumer Protection
Workbook Activity	68	Defining Community
Workbook Activity	69	Community Health Resources
Workbook Activity	70	Community Health Advocacy Skills
Workbook Activity	71	Health and the Environment
Workbook Activity	72	Air Pollution and Health
Workbook Activity	73	Water and Land Pollution and Health
Workbook Activity	74	Promoting a Healthy Environment

Name _____ Date _____ Period _____

Chapter 1
Workbook Activity
1

Cells, Tissues, and Organs

Directions Complete each sentence by writing the letter of the *best* word or words in the space on the left-hand side of the page.

_____ 1) Your body is made up of _____, or tiny living units.

_____ 2) A _____, or control center, can be found at the center of every cell.

_____ 3) Cells divide by a process called _____.

_____ 4) Cells that perform the same function cluster together in groups to form _____ such as muscle or bone.

_____ 5) The thin skin around the cell is called the _____.

_____ 6) The _____ is the jelly-like liquid floating inside the cell membrane.

_____ 7) A _____ is a group of organs working together to do a job.

_____ 8) Different kinds of tissues working together form an _____.

_____ 9) Many _____ kinds of cells make up your body.

_____ 10) One kind of cell inside your body is _____ cells.

a) body system
b) cells
c) cell membrane
d) cytoplasm
e) different
f) mitosis
g) nerve
h) nucleus
i) organ
j) tissue

Discover Health

Name	Date	Period	Chapter 1

Workbook Activity 2

The Body's Protective Covering

Directions Write *T* if the statement is true or *F* if it is false.

_____ 1) Your skin is made up of five layers.

_____ 2) Oil glands in the dermis help keep your skin smooth and stretchable.

_____ 3) Your skin is the second largest organ in the body.

_____ 4) Pores are tiny holes in your skin.

_____ 5) Oil in your skin helps keep your body cool.

_____ 6) The bottom layer of the skin is the subcutaneous layer.

_____ 7) The color of your skin and hair comes from keratin.

_____ 8) Melanin is a protein that hardens into nails.

_____ 9) The outer layer of your skin is called the dermis.

_____ 10) The dermis has blood vessels, nerves, and glands.

Discover Health

Name _____ Date _____ Period _____

Chapter 1
Workbook Activity
3

The Skeletal and Muscular Systems

Directions Complete each sentence by writing the letter of the *best* word or words in the space on the left-hand side of the page.

_____ 1) The skeleton is made up of _____ bones.

_____ 2) The bones of your _____ help protect body parts inside your body.

_____ 3) Without your skeleton, you would not be able to stand or _____ .

_____ 4) Bones move when muscles _____ .

_____ 5) Muscles always work in _____ .

_____ 6) You use _____ muscles when you want to walk.

_____ 7) Your heart muscle is a/an _____ muscle.

_____ 8) _____ connect your bones together.

_____ 9) When your muscles contract, they become _____ .

_____ 10) There are _____ kinds of muscles in your body.

a) contract
b) involuntary
c) joints
d) pairs
e) shorter
f) skeleton
g) two
h) 206
i) voluntary
j) walk

Discover Health

The Digestive and Excretory Systems

Directions Complete each sentence by writing the letter of the *best* word or words in the space on the left-hand side of the page.

_____ 1) The liquid in your mouth is called _____ .

_____ 2) Muscles force food through your _____ and into your stomach.

_____ 3) From the stomach, food moves into your _____ .

_____ 4) In the _____ , most of the water in undigested food returns to the blood.

_____ 5) Solid waste is stored in the _____ .

_____ 6) Solid waste passes out of the body through the _____ .

_____ 7) _____ is liquid waste that passes from your body.

_____ 8) Your digestive system consists of all the organs that break down _____ into substances the cells can use.

_____ 9) A _____ is a kind of tube through which urine passes.

_____ 10) Your body rids itself of wastes through the _____ system.

a) anus
b) esophagus
c) excretory
d) food
e) large intestine
f) rectum
g) saliva
h) small intestine
i) ureter
j) urine

Discover Health

Name _____ Date _____ Period _____ Chapter 1

Workbook Activity 5

The Respiratory and Circulatory Systems

Directions Complete each sentence by writing the letter of the *best* word or words in the space on the left-hand side of the page.

_____ 1) Tiny blood vessels are called _____.

_____ 2) There are two kinds of _____ in the blood.

_____ 3) The body's largest artery is the _____.

_____ 4) Blood vessels that carry blood to the heart are _____.

_____ 5) The heart is the main organ of the _____.

_____ 6) Blood vessels that carry blood away from the heart are _____.

_____ 7) Another name for the trachea is the _____.

_____ 8) When you breathe out, you get rid of _____.

_____ 9) The _____ both gets oxygen into the body and removes waste gas.

_____ 10) Your blood is made up of a liquid part called _____ and blood cells.

a) aorta
b) arteries
c) blood cells
d) capillaries
e) carbon dioxide
f) circulatory system
g) plasma
h) respiratory system
i) veins
j) windpipe

Discover Health

Name _____ Date _____ Period _____

Chapter 1
Workbook Activity
6

The Nervous System

Directions Complete each sentence by writing the letter of the *best* word or words in the space on the left-hand side of the page.

_____ 1) A thick bundle of nerves is the _____ .

_____ 2) Your brain is made up of _____ main parts.

_____ 3) The _____ has two halves.

_____ 4) The _____ works closely with the cerebrum.

_____ 5) The _____ controls involuntary muscles.

_____ 6) The brain and spinal cord make up the _____ nervous system.

_____ 7) Messages to and from the _____ travel at more than 100 miles an hour.

_____ 8) The brain's _____ hemisphere controls the ability to do math.

_____ 9) The _____ hemisphere in the brain controls your artistic ability.

_____ 10) The _____ nervous system includes all the nerves outside the central nervous system.

a) brain
b) central
c) cerebrum
d) cerebellum
e) left
f) medulla
g) peripheral
h) right
i) spinal cord
j) three

Discover Health

Name _____ Date _____ Period _____

Chapter 1
Workbook Activity
7

The Endocrine System

Directions Write *T* if the statement is true or *F* if it is false.

_____ 1) The endocrine system controls many skin cells.

_____ 2) Glands secrete chemicals.

_____ 3) Hormones are the same as body organs.

_____ 4) The pituitary gland is sometimes called the body's missing gland.

_____ 5) The thyroid gland is a large gland.

_____ 6) The adrenal glands secrete a chemical called adrenaline.

_____ 7) The pituitary gland is the largest gland in the body.

_____ 8) Hormones never enter the blood.

_____ 9) The endocrine system directs your body's growth.

_____ 10) The pituitary gland releases the hormones that cause the development of your reproductive system.

Discover Health

Name _____ Date _____ Period _____

Chapter 1
Workbook Activity
8

The Reproductive System

Directions Write *T* if the statement is true or *F* if it is false.

_____ 1) Sexual intercourse is the way a male sperm and female egg form new life.

_____ 2) The monthly removal of the lining of the uterus and the unfertilized egg in females is called mentimation.

_____ 3) In males the sex cells, known as sperm, are produced in the testes, the male sex glands.

_____ 4) The ovaries store and release the female's eggs.

_____ 5) Puberty is the time when children are born.

_____ 6) The vagina is the birth canal.

_____ 7) The temperature of the testes is lower than inside the body.

_____ 8) About four thousand sperm are released during sexual intercourse.

_____ 9) Inside the female, babies develop in the ureter.

_____ 10) The release of mature eggs every month is called ovulation.

Discover Health

Name _____ Date _____ Period _____

Chapter 2
Workbook Activity
9

Hygiene

Part A Directions Write *T* if the statement is true or *F* if it is false.

_____ 1) Persons who have oily skin should bathe less often than others.

_____ 2) It is better to allow your hair to dry naturally rather than to use a hair dryer.

_____ 3) Wait at least a half hour after exercising to wash up.

_____ 4) An orthodontist is a condition in which the eyeball lacks certain nerve endings called cones.

_____ 5) Periodontal disease is a condition of the endocrine glands.

_____ 6) Avoid wearing glasses or contact lenses if you can possibly help it.

Part B Directions Place a check mark next to the *best* answer.

7) Good skin care *does not* include:
_____ a) wearing a sunscreen.
_____ b) heavy exposure to cold and high winds.
_____ c) keeping your skin clean.
_____ d) the use of moisturizing creams or lotions.

8) An ophthalmologist is a doctor who specializes in:
_____ a) skin care.
_____ b) malocclusion.
_____ c) eye disease.
_____ d) tooth decay.

9) An orthodontist treats:
_____ a) an eye disease caused by a build-up of pressure in the eyeball.
_____ b) skin problems.
_____ c) the buildup of hard tartar or calculus on the fingernails.
_____ d) teeth that are crooked, crowded, and out of alignment.

10) Hearing loss *does not* occur as a result of:
_____ a) infection.
_____ b) the buildup of wax.
_____ c) caries.
_____ d) exposure to loud noises.

Discover Health

Name _____ Date _____ Period _____ | Chapter 2
Workbook Activity
10

Fitness

Directions Write *T* if the statement is true or *F* if it is false

_____ 1) Another word for "heart rate" is blood pressure.

_____ 2) Aerobic exercises are easiest to do while seated.

_____ 3) Stretching and bending exercises make you become dizzy.

_____ 4) Regular exercise can help everybody.

_____ 5) Most young people need less than six hours of sleep a night.

_____ 6) When you are tired or fatigued, your body actually functions better.

_____ 7) Vigorous exercise without first warming up can be harmful to your body.

_____ 8) People who have a risk of cardiovascular diseases should avoid exercising.

_____ 9) Often listening to truck traffic at 90 decibels could cause hearing loss.

_____ 10) High blood pressure can lead to heart attacks.

Discover Health

Name _____ Date _____ Period _____

Chapter 3
Workbook Activity
11

The Family Life Cycle

Directions Complete each sentence by writing the letter of the *best* word or words in the space on the left-hand side of the page.

_____ 1) _____ families include people from different generations.

_____ 2) _____ is a key to making a family work.

_____ 3) The changes in families over time are known as the family _____ .

_____ 4) Commitment to working things out and strong _____ skills are key elements in a healthy family.

_____ 5) Couples have new _____ when they become parents.

_____ 6) Parents must prepare adolescents to live _____ .

_____ 7) _____ each other is one of the key skills that help marriages work.

_____ 8) When children leave home and begin leading their own lives, _____ must learn to form new relationships with their children's spouses and in-laws.

_____ 9) As people grow older they must learn to _____ growing disabilities and limitations.

_____ 10) During the last stage of the family life cycle, a family member must learn to cope with the _____ of his or her own spouse.

a) accept
b) commitment
c) communication
d) death
e) extended
f) independently
g) life cycle
h) parents
i) supporting
j) responsibilities

Discover Health

Name _____ Date _____ Period _____

Chapter 3
Workbook Activity
12

Dealing With Family Problems

Part A Directions Place a check mark next to the only answer that *does not* apply.

1) To deal with family problems in a healthy way:
   _____ a) find support from a caring adult.
   _____ b) deny that a problem exists.
   _____ c) ask for help.
   _____ d) write about the problem in a diary.

2) Who can best help when a teenager is facing family problems?
   _____ a) parent, sibling, or other relative
   _____ b) a favorite school teacher
   _____ c) a codependent parent
   _____ d) a social worker or psychologist

3) The four stages of reacting to a problem include:
   _____ a) denial.
   _____ b) anger.
   _____ c) dysfunction.
   _____ d) acceptance.

4) Families handle stress by:
   _____ a) working together to find solutions.
   _____ b) setting reasonable guidelines.
   _____ c) getting a divorce.
   _____ d) communicating effectively.

5) A person may experience the four stages of grief during:
   _____ a) divorce.
   _____ b) loss of job.
   _____ c) illness.
   _____ d) a wedding.

Part B Directions Complete each sentence by writing the letter of the *best* word or words in the space on the left-hand side of the page.

_____ 6) When parents divorce, children may feel _____ .
_____ 7) Words and actions that _____ others are violent.
_____ 8) Examples of _____ are harsh words, threats, or lack of affection.
_____ 9) The most helpful way to react to problems is with a _____ attitude.
_____ 10) Writing feelings in a _____ can help you avoid depression.

a) depressed
b) diary
c) emotional abuse
d) hurt
e) positive

Discover Health

Name _____ Date _____ Period _____

Chapter 4
Workbook Activity
13

Emotions and Their Causes

Directions Complete each sentence by writing the letter of the *best* word or words in the space on the left-hand side of the page.

_____ 1) Your feelings are also called _____.

_____ 2) An emotion is made up of _____ parts.

_____ 3) Emotions may cause _____ changes in your body.

_____ 4) The mental side of emotions involves _____ why you feel a certain way.

_____ 5) We experience _____ when we feel threatened.

_____ 6) Fear can sometimes help you _____ yourself.

_____ 7) Deciding how to act when you are _____ is not always easy.

_____ 8) When you change the way you behave, you _____.

_____ 9) Relief and _____ are signs that a stressful situation is over.

_____ 10) Your feelings affect how you _____ to things.

a) adapt
b) angry
c) emotions
d) joy
e) physical
f) protect
g) react
h) stress
i) two
j) understanding

Discover Health

Name _____ Date _____ Period _____

Chapter 4
Workbook Activity
14

Social Emotions

Directions Complete each sentence by writing the letter of the *best* word or words in the space on the left-hand side of the page.

_____ 1) _____ are emotions that have to do with relationships.

_____ 2) You feel _____ when you disappoint yourself or others.

_____ 3) If someone does not approve of you, you might feel _____ .

_____ 4) Grief is a normal way to react to _____ .

_____ 5) If you lose a close relationship, you feel _____ .

_____ 6) Extreme sadness is also called _____ .

_____ 7) The best "cure" for depression is _____ .

_____ 8) Shame sometimes has outward signs like _____ .

_____ 9) The emotion of _____ is important for healthy living.

_____ 10) Signs of depression include _____ and feelings of anger.

a) blushing
b) depression
c) grief
d) guilty
e) loss
f) love
g) low energy
h) positive action
i) shame
j) social emotions

Discover Health

Name _____ Date _____ Period _____

Chapter 4
Workbook Activity
15

Emotions and Behavior

Directions Complete each sentence by writing the letter of the *best* word or words in the space on the left-hand side of the page.

_____ 1) You feel a _____ when you are missing something.

_____ 2) When you have a need, you can do _____ things: identify the cause, decide how to solve, and carry out the plan.

_____ 3) Behaving in a _____ way helps you feel better about yourself.

_____ 4) An agreement in which both sides give up a little is a _____ .

_____ 5) _____ are not "good" or "bad."

_____ 6) A good way to deal with emotions is to _____ them.

_____ 7) _____ is another good way to deal with emotions.

_____ 8) If you can't handle your emotions yourself, it's a good idea to talk to a _____ .

_____ 9) Thinking before you _____ is likely to help you behave more reasonably.

_____ 10) Hoping others guess what you need does not work. Being _____ works best.

a) act

b) compromise

c) direct

d) emotions

e) express

f) mental health professional

g) need

h) physical action

i) reasonable

j) three

Discover Health

Name _____ Date _____ Period _____ | Chapter 5
Workbook Activity
16

Managing Frustration

Part A Directions Write the letter of the item from Column B that matches the description in Column A.

Column A

_____ 1) whining when a person does not get his or her way

_____ 2) any strong and unfriendly action that intends harm

_____ 3) the unpleasant feeling you have when you are blocked from meeting your goals

_____ 4) giving up at the first sign of difficulty and refusing to continue to work on a problem

_____ 5) to handle a problem or get rid of the source of frustration

Column B

a) aggression

b) acting childish

c) cope

d) frustration

e) withdrawal

Part B Directions Write *T* if the statement is true or *F* if it is false.

_____ 6) People react to frustration the same way.

_____ 7) Acting childish or bullying others are good ways to get what you want.

_____ 8) One way to cope is by taking a second look at your goals.

_____ 9) To get over frustration, figure out what is blocking you.

_____ 10) If you work less hard, you will probably be able to reach your goals easier.

Discover Health

Name _____ Date _____ Period _____

Chapter 5 Workbook Activity 17

Managing Stress and Anxiety

Part A Directions Write the letter of the item from Column B that matches the description in Column A.

Column A   **Column B**

_____ 1) situations that seem to be dangerous   a) anxiety

_____ 2) how one feels about himself or herself   b) hassle

_____ 3) a feeling like fear with no clear reason   c) self-esteem

_____ 4) small, annoying event or problem   d) stress response

_____ 5) the body's reaction to stress   e) threats

Part B Directions Write T if the statement is true or F if it is false.

_____ 6) Major life changes are causes of stress.

_____ 7) A person's health is usually harmed by one hassle.

_____ 8) There are three parts to the stress response.

_____ 9) The fight-or-flight response is rare in people.

_____ 10) A poor way to manage stress is to prepare for it.

_____ 11) Avoiding your fear will often help you get over it.

_____ 12) Your body uses a lot of energy dealing with stress.

_____ 13) When you face stress, you may notice few physical signs.

_____ 14) Stress can be good or bad.

_____ 15) Your self-esteem is helped when someone puts you down.

Discover Health

Name _____ Date _____ Period _____

Chapter 5
Workbook Activity
18

Responding to Peer Pressure

Part A Directions Write the letter of the item from Column B that matches the description in Column A.

Column A

_____ 1) a person in the same age group as another
_____ 2) helpful kind of peer pressure
_____ 3) example of negative peer pressure
_____ 4) to act against
_____ 5) influence that your peers have on you

Column B

a) cheating because your friends want you to
b) peer
c) peer pressure
d) positive peer pressure
e) resist

Part B Directions Put a check mark by the letter that *best* completes each statement.

6) You *must* resist if people want you to:
_____ a) do something wrong or dangerous.
_____ b) go for a long walk.
_____ c) take a test.
_____ d) explain your reasons for refusing.

7) Sometimes, if you disagree with a group, they may:
_____ a) take you along to the game.
_____ b) become your friends.
_____ c) push you out of the group.
_____ d) try to use positive peer pressure with you.

8) Which of the following is *not* an example of peer pressure?
_____ a) The group expects you to pay attention to them.
_____ b) The group expects you to support them.
_____ c) The group expects you to wear similar clothes.
_____ d) The group expects you never to do what they want to do.

9) Finding someone in a group who agrees with you is a way to:
_____ a) resist peer pressure.
_____ b) find friends.
_____ c) push someone out of your group.
_____ d) all of the above.

10) To resist peer pressure, do *not*:
_____ a) ask questions.
_____ b) express your feelings.
_____ c) avoid others who agree with you.
_____ d) all of the above.

Discover Health

Name _____ Date _____ Period _____

Chapter 5
Workbook Activity
19

Eating Disorders

Part A Directions Complete each sentence by writing the letter of the best word or words from the Word Bank in the space on the left-hand side of the page.

Word Bank
a) image c) eating disorder e) bulimia
b) media d) anorexia

_____ 1) Body _____ is the way each individual sees himself or herself.

_____ 2) Teens often form harmful perceptions of their bodies because of the way the _____ show men and women.

_____ 3) An _____ causes people to lose control of their eating.

_____ 4) Usually, girls who have developed _____ choose not to eat because they think they are too fat.

_____ 5) The person with _____ is on a destructive cycle of eating and ridding the body of food.

Part B Directions Write *T* if the statement is true or *F* if it is false.

_____ 6) A person who overeats regularly should see a doctor.

_____ 7) Bulimia may cause tooth decay, rapid loss of weight, or death.

_____ 8) Anorexia results from emotional problems.

_____ 9) Anorexia affects girls more often than it does boys.

_____ 10) Most people's bodies are not like the ones they see in the media.

Discover Health

Name _____ Date _____ Period _____

Chapter 6
Workbook Activity
20

Being a Friend to Yourself

Directions Write *T* if the statement is true or *F* if it is false.

_____ 1) A feeling of emotional closeness to someone is called a bond.

_____ 2) Putting yourself down is a good way to reach your goals.

_____ 3) Positive self-talk is a way of talking nicely to yourself.

_____ 4) Taking care of yourself is a way to be good to yourself.

_____ 5) You should let others take care of your needs when possible.

_____ 6) People sometimes blame themselves for things that are not their fault.

_____ 7) Friends help you meet the need to belong.

_____ 8) Others need to be your friend before you can be your own friend.

_____ 9) Thinking about things you like about yourself is better than thinking about things you don't like.

_____ 10) Whenever you feel proud of something you have done, put yourself down.

Discover Health

Name	Date	Period	Chapter 6

Workbook Activity 21

Making Friends

Part A Directions Place a check mark next to the *best* answer.

1) Friends usually:
   _____ **a)** are exactly the same.
   _____ **b)** have things in common.
   _____ **c)** dislike each other.
   _____ **d)** have the same last name.

2) Talk that is interesting but not important is:
   _____ **a)** big talk.
   _____ **b)** important.
   _____ **c)** small talk.
   _____ **d)** useless.

3) People probably will be friendly to you if you:
   _____ **a)** are friendly to them.
   _____ **b)** ignore them.
   _____ **c)** call them names.
   _____ **d)** try to make them pay attention to you.

4) A way to be a good listener is:
   _____ **a)** to pay attention to what the other person says.
   _____ **b)** to look at the person when he or she talks.
   _____ **c)** ask questions.
   _____ **d)** all of the above

5) A way to treat a new friend nicely is to:
   _____ **a)** thank your friend for coming over.
   _____ **b)** make your friend pay attention to everything you do.
   _____ **c)** be sure your friend listens to your every word.
   _____ **d)** all of the above

Part B Directions Write *T* if the statement is true or *F* if it is false.

_____ 6) It's important to find friends who have different values from you.

_____ 7) If you feel all people should be treated equally, you should join a group that makes fun of others.

_____ 8) By sharing activities, you get to know another person.

_____ 9) Be sure not to respect your friend's rights.

_____ 10) True friends pressure others to do things their way.

Discover Health

Name _____ Date _____ Period _____

Chapter 6
Workbook Activity
22

Healthy Relationships

Directions Write *T* if the statement is true or *F* if it is false.

_____ 1) A relationship is a connection between people.

_____ 2) When you are uncomfortable around others, you are shy.

_____ 3) Gossip is positive rumors about someone.

_____ 4) When a relationship ends, be a friend to yourself.

_____ 5) Others might think a shy person is very friendly.

_____ 6) A healthy relationship respects your rights, needs, and values.

_____ 7) Friends need to do everything together.

_____ 8) A good relationship will allow you to be yourself.

_____ 9) Show your interest in others by waiting for them to ask lots of questions about you.

_____ 10) Spreading rumors about a person is a great way to make friends with that person.

Discover Health

Name	Date	Period	Chapter 7

Workbook Activity 23

Food for Energy

Directions Place a check mark next to the *best* answer.

1) What is the process of cells using nutrients called?
- ____ **a)** digestion
- ____ **b)** metabolism
- ____ **c)** absorption
- ____ **d)** all of the above

2) What is the process of breaking down food into nutrients called?
- ____ **a)** digestion
- ____ **b)** metabolism
- ____ **c)** absorption
- ____ **d)** all of the above

3) What is the process of moving nutrients into the bloodstream called?
- ____ **a)** digestion
- ____ **b)** metabolism
- ____ **c)** absorption
- ____ **d)** all of the above

4) Which is not a nutrient?
- ____ **a)** fat
- ____ **b)** calories
- ____ **c)** minerals
- ____ **d)** water

5) What starts the process of digestion?
- ____ **a)** chewing and the chemicals in saliva
- ____ **b)** stomach acids breaking down food
- ____ **c)** absorption into the large intestine
- ____ **d)** moving nutrients from the blood into your stomach

6) What does a calorie do?
- ____ **a)** absorbs energy in the body
- ____ **b)** helps digest food
- ____ **c)** moves nutrients from the stomach into the bloodstream
- ____ **d)** measures how much energy foods give the body

7) Where does metabolism occur?
- ____ **a)** in the stomach
- ____ **b)** in the lungs
- ____ **c)** in the small intestine
- ____ **d)** in the cells

8) Where does digestion occur?
- ____ **a)** in the small intestine
- ____ **b)** in the mouth
- ____ **c)** in the stomach
- ____ **d)** all of the above

9) What carries nutrients to the body's cells?
- ____ **a)** the circulatory system
- ____ **b)** the excretory system
- ____ **c)** the lungs
- ____ **d)** the brain

10) How many main steps are there to turn food into energy?
- ____ **a)** one
- ____ **b)** two
- ____ **c)** three
- ____ **d)** four

Discover Health

Name _____ Date _____ Period _____ | Chapter 7
Workbook Activity 24

Carbohydrates and Protein

Directions Write *T* if the statement is true or *F* if it is false.

_____ 1) An example of a simple carbohydrate is cookies.

_____ 2) Amino acids are building blocks in protein.

_____ 3) You should get 12 to 15 percent of your calories from protein.

_____ 4) Only foods from animal sources provide complete proteins.

_____ 5) Complex carbohydrates such as vegetables and whole grains are a source of fiber.

_____ 6) Fiber is the part of food that is easiest to digest.

_____ 7) A complete protein contains all twenty-two proteins.

_____ 8) Foods from only one plant contain complete proteins.

_____ 9) Fiber is not a nutrient.

_____ 10) Six essential nutrients are needed for people to live.

Discover Health

Name _____ Date _____ Period _____

Chapter 7
Workbook Activity
25

Fats and Cholesterol

Directions Write *T* if the statement is true or *F* if it is false.

_____ 1) Too much fat can lead to heart disease.

_____ 2) Polyunsaturated fats usually come from plants.

_____ 3) Cholesterol comes only from animal foods.

_____ 4) Cholesterol can be good or bad.

_____ 5) Protein is a waxy, fat-like substance found in every cell in the body.

_____ 6) Salad dressings, dips, and gravies are part of a low-fat diet.

_____ 7) More than a third of your daily calories should come from fat.

_____ 8) There are three main kinds of fats.

_____ 9) Cholesterol can build up in your blood vessels.

_____ 10) Your body needs some cholesterol to stay healthy.

Discover Health

Name _____ Date _____ Period _____

Chapter 7
Workbook Activity
26

Vitamins, Minerals, and Water

Directions Place a check mark next to the *best* answer.

1) Which are fat-soluble vitamins?
   ____ **a)** vitamin A
   ____ **b)** vitamin D
   ____ **c)** vitamin K
   ____ **d)** all of the above

2) Where are minerals formed?
   ____ **a)** in trees
   ____ **b)** in the earth
   ____ **c)** in grass
   ____ **d)** in animal products

3) Boys and girls need how much calcium?
   ____ **a)** boys need more calcium
   ____ **b)** girls need more calcium
   ____ **c)** boys and girls need about the same
   ____ **d)** none of the above

4) How much of your body weight is water?
   ____ **a)** one-third
   ____ **b)** a half
   ____ **c)** 60 percent
   ____ **d)** 90 percent

5) Sodium is important for what bodily function?
   ____ **a)** helps nerve cells communicate
   ____ **b)** causes brain cells to grow
   ____ **c)** helps digestion
   ____ **d)** is not important for anything

6) Which mineral helps bones and teeth?
   ____ **a)** calcium
   ____ **b)** nitrogen
   ____ **c)** sodium
   ____ **d)** iron

7) Which of the following is not an essential nutrient?
   ____ **a)** water
   ____ **b)** vitamins
   ____ **c)** minerals
   ____ **d)** fiber

8) Which vitamin dissolves in water?
   ____ **a)** vitamin A
   ____ **b)** vitamin B
   ____ **c)** vitamin E
   ____ **d)** vitamin D

9) What is the best way to preserve vitamins B and C in cooking?
   ____ **a)** let foods soak in water
   ____ **b)** wash foods too long
   ____ **c)** soak foods for a short time
   ____ **d)** all of the above

10) What is the best way to get a variety of vitamins?
   ____ **a)** eat a wide variety of foods
   ____ **b)** eat mainly animal products
   ____ **c)** take dietary supplements
   ____ **d)** choose foods high in fiber

Discover Health

Name _____ Date _____ Period _____

Chapter 7
Workbook Activity
27

Dietary Guidelines

Part A Directions Write *T* if the statement is true or *F* if it is false.

_____ 1) Teenage girls need more iron than teenage boys require.

_____ 2) Choosing a diet low in sugar is part of a healthy diet.

_____ 3) You should always choose the smaller serving on the Food Guide Pyramid.

_____ 4) Breads and grains form the base of the Food Guide Pyramid.

_____ 5) The National Food Service (NSF) created the Food Guide Pyramid.

Part B Directions Place a check mark next to the *best* answer.

6) A chart that can be used to choose a healthy diet is:
_____ a) the Food Guide Pyramid.
_____ b) the Recommended Daily Allowance chart.
_____ c) the Percent of Daily Diet chart.
_____ d) the Nutrition Facts label.

7) What is a serving size?
_____ a) 16 ounces or more
_____ b) a way to measure how much of a food should be eaten
_____ c) a measure of the fat in a serving of food
_____ d) a way to measure the calories in a serving of food

8) Who needs more calories, calcium, and iron?
_____ a) parents
_____ b) teenagers
_____ c) grandparents
_____ d) active businesspeople

9) Which of these is a general dietary guideline?
_____ a) Eat a wide variety of foods.
_____ b) Balance the foods you eat with physical activity.
_____ c) Eat many grains, fruits, and vegetables every day.
_____ d) all of the above

10) Different people need different amounts of foods depending on:
_____ a) age.
_____ b) activity level.
_____ c) gender.
_____ d) all of the above

Discover Health

Name _____ Date _____ Period _____

Chapter 8
Workbook Activity
28

Food Choices and Health

Directions Complete each sentence by writing the letter of the *best* word or words from the Word Bank in the space on the left-hand side of the page.

Word Bank

a) anorexia nervosa   f) obesity
b) bloated   g) one-third
c) bulimia   h) malnutrition
d) calories   i) raw vegetables
e) an eating disorder   j) snack

_____ 1) Boys need more _____ than girls beginning at age 11 and continuing throughout life.

_____ 2) It's okay for teens to _____ because they have high energy needs.

_____ 3) _____ occurs when the body does not get enough calories or nutrients.

_____ 4) _____ is an eating disorder in which one avoids eating.

_____ 5) A pattern of eating that leads to health problems is called _____.

_____ 6) _____ is a condition in which a person is more than 20 percent overweight.

_____ 7) Malnutrition can cause _____, or swollen, stomachs.

_____ 8) People with _____ often eat large amounts of food and then vomit to rid themselves of it.

_____ 9) About _____ of teens' calories come from snacks.

_____ 10) A good snack choice is _____.

Discover Health

Name _____ Date _____ Period _____

Chapter 8
Workbook Activity
29

Influences on Food Choices

Part A Directions Place a check mark beside the one answer that *does not* apply.

1) What factors influence the foods you choose to eat?
   _____ a) taste
   _____ b) availability
   _____ c) weather
   _____ d) messages on TV

2) How do feelings affect your food choices?
   _____ a) You choose to eat foods that make you feel good.
   _____ b) Food can improve your emotional well-being.
   _____ c) A pleasant eating experience can mask feelings of anger or frustration.
   _____ d) Food may be used as a reward or punishment.

3) How does your environment affect your food choices?
   _____ a) You eat what your parents provide.
   _____ b) You choose restaurant foods from the menu.
   _____ c) You try what your friends eat.
   _____ d) You select the healthiest candy bar by reading the label.

4) How do family and friends affect your food choices, according to the text?
   _____ a) You often choose foods your friends dislike as a way to irritate them.
   _____ b) You choose to eat certain foods to be accepted by your friends.
   _____ c) You eat a certain food because you don't want to hurt someone's feelings.
   _____ d) You enjoy traditional foods from your own culture.

5) What do advertisers do to attract children to their products?
   _____ a) use colorful packaging
   _____ b) display the products on bottom shelves
   _____ c) point out the nutritional advantages of their products
   _____ d) emphasize how much fun the food is to eat

Part B Directions Write *T* if the statement is true or *F* if it is false.

_____ 6) You make some food choices based on social reasons.

_____ 7) The foods on a restaurant's menu may limit your healthy choices.

_____ 8) Food is rarely used as a reward or a punishment.

_____ 9) Food packaging is meant to attract a person's attention.

_____ 10) You may learn about another culture by eating its food.

Discover Health

Name _____ Date _____ Period _____

Chapter 8
Workbook Activity
30

Reading Food Labels

Part A Directions Place a check mark next to the *best* answer.

1) Which items must appear on a food label?
   _____ a) name of product, weight, list of ingredients, manufacturer's name and address
   _____ b) name of product, volume, list of ingredients, number of calories
   _____ c) name of product, volume, manufacturer's name and address, number of calories
   _____ d) manufacturer's name and address, weight, list of ingredients, number of calories

2) If the second ingredient listed on a product label is sugar, this tells you:
   _____ a) the product contains too much sugar.
   _____ b) the product is not healthy.
   _____ c) sugar is the ingredient present in the second largest quantity.
   _____ d) sugar is an additive.

3) The list of ingredients is important for people who:
   _____ a) have food allergies.
   _____ b) need organized lists.
   _____ c) eat 2,500 calories per day.
   _____ d) already know how many nutrients they get in food.

4) One way to make healthy food choices is by:
   _____ a) comparing prices.
   _____ b) reading the information on food labels.
   _____ c) comparing servings on cans with servings in restaurants.
   _____ d) paying attention to television ads.

5) What is information you can gain from reading labels?
   _____ a) serving size
   _____ b) calories per serving
   _____ c) amounts of different vitamins and minerals
   _____ d) all of the above

Part B Directions Write the letter of the item from Column B that matches the description in Column A.

Column A

_____ 6) a number that identifies a group of packages

_____ 7) the part of a label that tells about the calories and nutrients in the food

_____ 8) the part of the package that tells what is in the package

_____ 9) amounts of nutrients used as standards on food labels

_____ 10) A percent of nutrients based on 2,000 calories per day

Column B

a) Daily Values

b) label

c) lot number

d) Nutrition Facts

e) U.S. Recommended Daily Allowances (U.S. RDA)

Discover Health

Name _____ Date _____ Period _____

| Chapter 8 |
| Workbook Activity |
| 31 |

Making Foods Safe

Directions Write the letter of the item from Column B that matches the description in Column A.

Column A

_____ 1) infect by contact with germs and poisons

_____ 2) the government agency responsible for regulating what information is required on food label

_____ 3) substances added to foods to make them better

_____ 4) additives that stop harmful germs from growing

_____ 5) a process during which nutrients are added to foods lost during processing

_____ 6) adding nutrients that are not normally found in a food to improve nutrition

_____ 7) an agency that grades products for quality and inspects meat during storage

_____ 8) can quickly grow on foods and cause food poisoning

_____ 9) occurs when food is handled by people who are sick, have open wounds on their hands, or have not washed their hands

_____ 10) example of a commonly fortified food

Column B

a) additives

b) contaminate

c) enrichment

d) fortification

e) food contamination

f) germs

g) milk

h) preservatives

i) U.S. Department of Agriculture (USDA)

j) U.S. Food and Drug Administration (FDA)

Discover Health

Name _____ Date _____ Period _____

Chapter 9
Workbook Activity
32

Prescription and Over-the-Counter Medicines

Part A Directions Write the letter of the item from Column B that matches the description in Column A.

Column A

_____ 1) medicines that can be bought in a store without a doctor's written order

_____ 2) drugs that can only be purchased with a written order from a doctor

_____ 3) drugs that treat allergy symptoms

_____ 4) substance that changes how the mind or body works

_____ 5) person trained and licensed to prepare prescription drugs

Column B

a) antihistamines

b) drug

c) over-the-counter medicines

d) pharmacist

e) prescription medicines

Part B Directions Complete each sentence by writing the letter of the *best* word or words in the space on the left-hand side of the page.

_____ 6) A drug used to prevent a disease or health problem is a _____.

_____ 7) Over-the-counter medicines can treat symptoms and relieve pain but do not _____ disease.

_____ 8) If over-the-counter drugs are used in the _____ way, they can make you sick.

_____ 9) Over-the-counter drugs can _____ the symptoms of a serious illness.

_____ 10) A medicine for minor skin infections is an _____.

f) cover up

g) cure

h) medicine

i) ointment

j) wrong

Discover Health

Name _____ Date _____ Period _____

Chapter 9
Workbook Activity
33

The Effect of Medicines and Drugs on the Body

Part A Directions Write the letter of the item from Column B that matches the description in Column A.

Column A

_____ 1) a way to absorb medicines through the skin

_____ 2) used to administer drugs through the rectum

_____ 3) fastest way to get drugs into the body

_____ 4) an unexpected result of taking medicine

_____ 5) a helpful result of taking medicine

Column B

a) injection

b) patch

c) suppositories

d) side effect

e) therapeutic effect

Part B Directions Complete each sentence by writing the letter of the *best* word or words in the space on the left-hand side of the page.

_____ 6) Never use a medicine without the knowledge and supervision of a _____ adult.

_____ 7) When using an over-the-counter medicine, read the _____ carefully and follow them exactly as they are stated.

_____ 8) Never mix _____ types of medicines at the same time unless the doctor has prescribed them for you.

_____ 9) _____ and other drugs should never be mixed.

_____ 10) Drugs can enter the body of an unborn _____ .

f) alcohol

g) baby

h) different

i) directions

j) trusted

Discover Health

Name _____ Date _____ Period _____ | Chapter 9
Workbook Activity 34

Tobacco

Directions Write *T* if the statement is true or *F* if it is false.

_____ 1) Drugs can be introduced into the body through smoking.

_____ 2) Nicotine is a stimulant found in tobacco.

_____ 3) A big reason why young people start smoking is peer pressure.

_____ 4) Smoking can cause many serious diseases.

_____ 5) Smokeless tobacco is tobacco that is put in a pipe and smoked.

_____ 6) Second-hand smoke is harmful to smokers only.

_____ 7) When people stop smoking, they might have a period of withdrawal.

_____ 8) Most flights in the United States allow smoking in the rest rooms.

_____ 9) Very few harmful chemicals are produced by burning tobacco.

_____ 10) Emphysema is a form of lung cancer.

_____ 11) Smoking can shorten a person's life.

_____ 12) Smoking is harmful, but smokeless tobacco is safe to use.

_____ 13) Skin patches or special gum can help a person stop smoking.

_____ 14) In the United States, it is hard to avoid breathing cigarette smoke.

_____ 15) There are laws against smoking in many public places.

Discover Health

Name	Date	Period	Chapter 9

Workbook Activity 35

Alcohol

Directions Write the letter of the item from Column B that matches the description in Column A.

Column A

_____ 1) a drink that contains alcohol

_____ 2) a chemical used to prevent the spread of disease

_____ 3) a drug that slows the central nervous system

_____ 4) drinking too much or too frequently

_____ 5) excited or stimulated by a drug

_____ 6) a disease in which a person is dependent on alcohol use

_____ 7) a kind of alcohol found in wine, beer, and hard liquors

_____ 8) a person in a group who will not drink and will drive the group home

_____ 9) an organization that helps people live alcohol-free lives

_____ 10) a reason why young people may use alcohol

Column B

a) alcohol abuse

b) alcoholic beverage

c) Alcoholics Anonymous

d) alcoholism

e) depressant

f) designated driver

g) disinfectant

h) ethyl alcohol

i) intoxicated

j) peer pressure

Discover Health

Name _____ Date _____ Period _____

Chapter 9
Workbook Activity
36

Narcotics, Depressants, Stimulants, and Hallucinogens

Directions Write the letter of the item from Column B that matches the description in Column A.

Column A

_____ 1) a drug that confuses the way the brain processes information

_____ 2) a synthetic stimulant

_____ 3) a dangerous and illegal narcotic

_____ 4) a stimulant found in coffee, tea, chocolate, and some soft drinks

_____ 5) a drug that dulls the senses or relieves pain

_____ 6) a dangerous and illegal stimulant made from the coca plant

_____ 7) a drug made from the opium poppy

_____ 8) a form of cocaine that is smoked

_____ 9) a narcotic made in laboratories

_____ 10) a form of an amphetamine that is smoked

_____ 11) a condition in which a person needs more and more of a drug for the same effect

_____ 12) a distortion of the senses caused by mental disease or drugs

_____ 13) a drawing tightly together and relaxing of a muscle

_____ 14) another term for hallucinogen

_____ 15) severe shaking

Column B

a) amphetamine

b) caffeine

c) cocaine

d) convulsion

e) crack cocaine

f) hallucination

g) hallucinogen

h) heroin

i) ice

j) narcotic

k) opiate

l) psychedelic drug

m) synthetic drug

n) tolerance

o) tremor

Discover Health

Name _____ Date _____ Period _____

Chapter 9
Workbook Activity
37

Other Dangerous Drugs

Directions Place a check mark beside the *best* answer.

1) Marijuana:
   _____ **a)** is an illegal drug. _____ **b)** can cause cancer.
   _____ **c)** is smoked, eaten, or drunk as tea. _____ **d)** all of the above

2) Marijuana:
   _____ **a)** can cause physical side effects. _____ **b)** interferes with memory.
   _____ **c)** takes away a person's energy. _____ **d)** all of the above

3) Breathable chemicals found in cleaners, paint removers, correction fluids, and gasoline are called:
   _____ **a)** inhalants. _____ **b)** designer drugs.
   _____ **c)** look-alike drugs. _____ **d)** anabolic steroids.

4) Drugs that are manufactured to have the same effects as illegal drugs are called:
   _____ **a)** inhalants. _____ **b)** designer drugs.
   _____ **c)** look-alike drugs. _____ **d)** anabolic steroids.

5) Chemicals that mimic the effects of the male hormone testosterone are called:
   _____ **a)** inhalants. _____ **b)** designer drugs.
   _____ **c)** look-alike drugs. _____ **d)** anabolic steroids.

6) Being sterile means:
   _____ **a)** unable to have children. _____ **b)** unable to hear.
   _____ **c)** unable to grow marijuana. _____ **d)** having hallucinations.

7) A steroid is:
   _____ **a)** a musical instrument. _____ **b)** a drug that causes death.
   _____ **c)** a sterile athlete. _____ **d)** a chemical that occurs naturally in the body.

8) Illegal manufactured drugs that are almost the same as legal drugs are:
   _____ **a)** inhalants. _____ **b)** designer drugs.
   _____ **c)** look-alike drugs. _____ **d)** anabolic steroids.

9) These drugs often contain dangerously large amounts of caffeine:
   _____ **a)** inhalants. _____ **b)** designer drugs.
   _____ **c)** look-alike drugs. _____ **d)** anabolic steroids.

10) Using inhalants can cause:
   _____ **a)** slurred speech and headaches. _____ **b)** permanent brain damage.
   _____ **c)** poor judgment and confusion. _____ **d)** all of the above

Discover Health

Name _____ Date _____ Period _____

Chapter 10
Workbook Activity
38

The Problems of Drug Dependence

Part A Directions Write the letter of the item from Column B that matches the description in Column A.

Column A

_____ 1) a disease spread by sexual contact

_____ 2) occurs through improper use of legal or illegal drugs

_____ 3) occurs when a drug-dependent person suddenly stops using the drug

_____ 4) the body's need to take more and more of a drug to get the same effect

_____ 5) the need for a drug resulting from frequent drug use

Column B

a) drug abuse

b) drug dependence

c) sexually transmitted disease

d) tolerance

e) withdrawal symptoms

Part B Directions Write *T* if the statement is true or *F* if it is false.

_____ 6) Drug abuse costs the United States trillions of dollars every year.

_____ 7) Drug dependence often has three stages.

_____ 8) Signs of a drug problem include loss of memory and trouble with the police.

_____ 9) A person trying to stop using drugs should be under a doctor's care.

_____ 10) Drug dependence causes many health, safety, and family problems.

Discover Health

Name _____ Date _____ Period _____

Chapter 10
Workbook Activity
39

Solutions to Drug Dependence

Directions Choose the *best* word or words from the Word Bank to match the descriptions below. Write the letter in the space on the left-hand side of the page.

Word Bank

a) admitting to a problem with drugs f) residential treatment center
b) Al-Anon g) saying no
c) Alateen h) sponsor
d) detoxification i) support group
e) outpatient treatment center j) three

_____ 1) The removal of a drug from the body

_____ 2) A support group for teens from alcoholic homes

_____ 3) A recovering alcoholic who helps others recover

_____ 4) A group of people with similar problems who help each other

_____ 5) A place where recovering drug users live for a time

_____ 6) Recovering drug users go here for part of each day for treatment.

_____ 7) This can help a person stay away from drugs.

_____ 8) A support group for adults dealing with a person with alcoholism

_____ 9) This is the first step to living without drugs.

_____ 10) The number of steps there are to recovering from drug dependence

Discover Health

Name _____ Date _____ Period _____ | **Chapter 11**
Workbook Activity
40

Causes of Disease

Directions Write the letter of the item from Column B that matches the description in Column A.

Column A

_____ 1) disease is caused by infection

_____ 2) disorder of normal body function

_____ 3) sticky fluid produced by mucous membranes

_____ 4) disease caused by a pathogen

_____ 5) sickness caused by a pathogen in the body

_____ 6) parts of a cell passed from parent to child

_____ 7) disease passed through genes

_____ 8) inherited disease in which the muscles do not develop normally

_____ 9) a disease-causing germ

_____ 10) thin, moist tissue that lines body openings

Column B

a) acquired disease

b) disease

c) gene

d) infection

e) infectious disease

f) inherited disease

g) muscular dystrophy

h) mucus

i) mucous membrane

j) pathogen

Discover Health

The Body's Protection From Disease

Chapter 11
Workbook Activity 41

Directions Write the letter of the item from Column B that matches the description in Column A.

Column A

_____ 1) injection of dead or weakened viruses

_____ 2) system of organs, tissues, and cells that fight infection

_____ 3) type of vaccination

_____ 4) body's first response to a pathogen

_____ 5) contains antibodies that protect babies from pathogens

_____ 6) protein that kills a specific pathogen

_____ 7) immune system

_____ 8) resistant to infection

_____ 9) part of the inflammatory response

_____ 10) means of making a body immune from a disease

Column B

a) antibody

b) body's final line of defense

c) injection

d) inflammatory response

e) immune

f) immune system

g) immunization

h) mother's milk

i) swelling blood vessels

j) vaccination

Discover Health

Name _____ Date _____ Period _____

Chapter 12
Workbook Activity
42

AIDS

Directions Match each item on the left with the correct item on the right. Write the correct letter on each blank. Each item on the right may be used once, more than once, or not at all.

_____ 1) Body system that fights disease

_____ 2) Rare type of cancer of the skin or internal organs

_____ 3) Having HIV in the blood

_____ 4) Virus that causes AIDS

_____ 5) Not having HIV in the blood

_____ 6) Transfer of blood from one to another

_____ 7) Able to be passed from one person to another

_____ 8) Acquired immunodeficiency syndrome

_____ 9) One way AIDS is spread

_____ 10) How AIDS is detected

a) AIDS

b) blood test

c) communicable

d) HIV

e) HIV-negative

f) HIV-positive

g) immune system

h) Kaposi's sarcoma

i) sexual contact

j) transfusion

Discover Health

Name _____ Date _____ Period _____

Chapter 12
Workbook Activity
43

Sexually Transmitted Diseases

Part A Directions Place a check mark beside *all* answers that apply.

1) Which are sexually transmitted diseases?
   _____ a) herpes, AIDS, gonorrhea, syphilis, chlamydia
   _____ b) gonorrhea, syphilis, chlamydia, chancre, penicillin
   _____ c) syphilis, chlamydia, trachoma, AIDS, chancre
   _____ d) trachoma, chancre, penicillin, herpes, AIDS

2) Gonorrhea:
   _____ a) is the most commonly reported infectious disease in the U.S.
   _____ b) if left untreated, progresses to three stages.
   _____ c) can cause swelling and redness in the genitals.
   _____ d) is slow to develop and difficult to diagnose.

3) Syphilis:
   _____ a) is the most commonly reported infectious disease in the U.S.
   _____ b) if left untreated, progresses to three stages.
   _____ c) can be spread from pregnant woman to fetus.
   _____ d) is slow to develop and difficult to diagnose.

4) STDs that are easily cured are:
   _____ a) chlamydia, herpes, gonorrhea.
   _____ b) chlamydia, gonorrhea, syphilis.
   _____ c) AIDS, gonorrhea, syphilis.
   _____ d) herpes, AIDS, syphilis.

Part B Directions Write *T* if the statement is true or *F* if it is false.

_____ 5) Genital herpes is a chronic disease.
_____ 6) AIDS is a sexually transmitted disease.
_____ 7) Herpes is easily cured with penicillin.
_____ 8) A small, painless, hard sore that appears on a person infected with syphilis.
_____ 9) Painful blisters on a person's genitals may indicate an active case of herpes.
_____ 10) Males infected with gonorrhea can become sterile if the disease is not treated promptly.
_____ 11) Gonorrhea can cause an eye infection in babies.
_____ 12) If a person acquires one STD, he or she is immune to other STDs.
_____ 13) Physicians have no treatment to offer patients with AIDS or herpes.
_____ 14) Chlamydia and gonorrhea have similar symptoms.
_____ 15) Blood tests before marriage can help detect STDs.

Discover Health

Name _____ Date _____ Period _____ | Chapter 13
Workbook Activity
44

Cardiovascular Diseases and Disorders

Part A Directions Write the letter of the item from Column B that *best* matches the description in Column A.

Column A

_____ 1) diseases of the heart and blood vessels

_____ 2) a thickening of the walls of the arteries

_____ 3) a condition in which fat builds up in the large arteries, causing them to become too narrow

_____ 4) the force of blood against the walls of your arteries when the heart pumps

_____ 5) another name for high blood pressure

_____ 6) condition in which blood to the brain is suddenly stopped

_____ 7) a habit or trait that is known to increase a person's chance of having a disease

_____ 8) occurs when the supply of blood or nutrients to the heart is cut off

Column B

a) atherosclerosis

b) blood pressure

c) cardiovascular disease

d) arteriosclerosis

e) hypertension

f) risk factor

g) stroke

h) heart attack

Part B Directions Write *T* if the statement is true or *F* if it is false.

_____ 9) Cardiovascular diseases are the leading cause of death in the United States.

_____ 10) At a person's birth, the walls of the arteries are rough and hard.

Discover Health

Name	Date	Period	Chapter 13
			Workbook Activity 45

Cancer

Directions Write the letter of the item from Column B that matches the description in Column A.

Column A

_____ 1) a large group of diseases marked by an abnormal and harmful growth of cells

_____ 2) a word used to describe harmful cancer cells

_____ 3) a mass of cells

_____ 4) using energy waves to destroy cancer

_____ 5) most common cancer in men older than age 65

_____ 6) cancer that has spread to distant sites in the body

_____ 7) most common cancer in the United States

_____ 8) one cancer that can be prevented

_____ 9) a word used to describe abnormal but harmless cells that are not cancerous

_____ 10) a form of drug treatment that destroys cancer cells

Column B

a) benign

b) cancer

c) chemotherapy

d) lung cancer

e) malignant

f) metastasized cancer

g) prostate cancer

h) radiation

i) skin cancer

j) tumor

Discover Health

Name _____ Date _____ Period _____

Chapter 13
Workbook Activity
46

Asthma

Directions Write *T* if the statement is true or *F* if it is false.

_____ 1) Asthma is a disease that attacks the heart.

_____ 2) There are fewer than five triggers for asthma.

_____ 3) One trigger of asthma can be strenuous exercise.

_____ 4) During an asthma attack, the airways in the nose and mouth become swollen.

_____ 5) Asthma commonly affects adults over the age of 21.

_____ 6) Asthma affects different people differently.

_____ 7) Asthma can be life threatening if left untreated.

_____ 8) Asthma causes mucus to be produced in the lungs.

_____ 9) People with asthma should avoid pollen and dust.

_____ 10) Asthma is caused only by nonfood items.

Discover Health

Name _____ Date _____ Period _____

| Chapter 13 |
| Workbook Activity |
| 47 |

Diabetes

Directions Write the letter of the item from Column B that matches the description in Column A.

Column A

_____ 1) first recognized during pregnancy and no longer present afterwards

_____ 2) a group of conditions that cause a high level of sugar in the blood

_____ 3) an important hormone, necessary to digest food

_____ 4) the name for the chemical source of sugar your body uses to give you energy

_____ 5) clouding of the lens of the eye

_____ 6) eye disease caused by pressure inside the eye

_____ 7) may cause amputation of hands or feet

_____ 8) may be controlled by diet, exercise, and insulin

_____ 9) may be caused by obesity and run in families

_____ 10) diabetes results when this gland stops making insulin

Column B

a) cataracts

b) diabetes

c) gestational diabetes

d) glaucoma

e) glucose

f) insulin

g) pancreas

h) severe nerve damage

i) Type I diabetes

j) Type II diabetes

Discover Health

Name _____ Date _____ Period _____

Chapter 13
Workbook Activity
48

Arthritis

Part A Directions Put a check mark next to the *best* answer.

1) Arthritis can be caused by:
_____ **a)** aging.
_____ **b)** germs.
_____ **c)** a defect in the immune system.
_____ **d)** all of the above

2) A chronic disease:
_____ **a)** lasts a long time.
_____ **b)** causes swelling of the joints.
_____ **c)** causes a person's joints to break down with old age.
_____ **d)** is caused by a defect in a person's immune system.

3) Septic arthritis:
_____ **a)** lasts a long time.
_____ **b)** causes swelling of the joints.
_____ **c)** causes a person's joints to break down with old age.
_____ **d)** is caused by a defect in a person's immune system.

4) Rheumatoid arthritis:
_____ **a)** lasts a long time.
_____ **b)** causes swelling of the joints.
_____ **c)** causes a person's joints to break down with old age.
_____ **d)** is caused by a defect in a person's immune system.

5) Osteoarthritis:
_____ **a)** lasts a long time.
_____ **b)** causes swelling of the joints.
_____ **c)** causes a person's joints to break down with old age.
_____ **d)** is caused by a defect in a person's immune system.

Part B Directions Write *T* if the statement is true or *F* if it is false.

_____ **6)** Chronic diseases can be managed but not cured.

_____ **7)** Rheumatoid arthritis occurs much more often in women than in men.

_____ **8)** Bacteria and viruses can cause joints to swell and fill with pus.

_____ **9)** In the first stage of osteoarthritis, the joints become soft.

_____ **10)** Pain relievers and physical therapy can help in the treatment of rheumatoid arthritis.

Discover Health

Name _____ Date _____ Period _____ | Chapter 13
Workbook Activity 49

Epilepsy

Part A Directions Write the letter of the item from Column B that matches the description in Column A.

Column A

_____ 1) a chronic disease that is caused by the malfunction of brain cells

_____ 2) physical and mental reactions to the brain's malfunction

_____ 3) number of kinds of epileptic seizures

_____ 4) only seizures longer than this are considered dangerous

_____ 5) a person stares for a moment or drops something

Column B

a) two

b) half an hour

c) petit mal seizure

d) epilepsy

e) seizures

Part B Directions Write *T* if the statement is true or *F* if it is false.

_____ 6) A person's muscles stiffen during a petit mal seizure.

_____ 7) Petit mal seizures begin in the part of the brain that controls thinking.

_____ 8) Grand mal seizures begin in the part of the brain that controls muscles.

_____ 9) Epilepsy occurs more often in men than in women.

_____ 10) Grand mal seizures usually cause brain damage.

_____ 11) A diagnosis of epilepsy is often made only when a person has had several seizures.

_____ 12) Blood tests can help confirm the disease of epilepsy.

_____ 13) The best treatment for epilepsy is brain surgery.

_____ 14) Most seizures have no known cause.

_____ 15) Febrile seizures usually lead to epilepsy later in life.

Discover Health

Name _____ Date _____ Period _____

| Chapter 14 |
| Workbook Activity |
| 50 |

Promoting Safety

Directions Write *T* if the statement is true or *F* if it is false.

_____ 1) When a person uses alcohol or other drugs, the risk of injury decreases.

_____ 2) Stimulants speed up the function of brain cells.

_____ 3) When in a car, keep noise levels down so the driver can concentrate.

_____ 4) Wear a seat belt only if you are the driver of a car.

_____ 5) Avoiding sports-related injuries is impossible.

_____ 6) Always wear the right safety equipment when playing sports.

_____ 7) A firearm is a handgun or a rifle.

_____ 8) Most gun accidents among young people occur because of curiosity about firearms.

_____ 9) Be sure to keep guns and ammunition locked together in a safe cabinet.

_____ 10) Riding with a driver who has been drinking is dangerous.

Discover Health

Reducing Risks of Fire

Chapter 14
Workbook Activity 51

Directions Write *T* if the statement is true or *F* if it is false.

_____ 1) Almost all house fires are preventable.

_____ 2) Test smoke detectors at home twice a year.

_____ 3) Store flammable materials in the hottest part of your basement.

_____ 4) A fire increases five hundred times every minute.

_____ 5) When leaving a burning building, try to take as many belongings as possible.

_____ 6) Fires usually burn for two to four minutes before a smoke detector goes off.

_____ 7) To avoid fire, don't overload electrical outlets.

_____ 8) Only let children play with safety matches.

_____ 9) There are a few situations in which it is safe to try to put out a fire.

_____ 10) If your home is on fire, go to a neighbor and call 911.

Discover Health

Name _____ Date _____ Period _____

Chapter 14
Workbook Activity
52

Safety for Teens

Directions Complete each sentence by writing the letter of the *best* word or words from the Word Bank in the space on the left-hand side of the page.

Word Bank

a) 911
b) address
c) door
d) E-mail
e) five

f) Internet
g) life-threatening emergency
h) parents
i) strangers
j) unattended

_____ 1) The _____ is a worldwide computer network.

_____ 2) When you baby-sit, never leave children _____ for even a short time.

_____ 3) In a _____, a person could die if medical treatment is not immediate.

_____ 4) When you call _____ speak clearly and don't hang up until told to.

_____ 5) Never accept rides from _____ even if they seem friendly.

_____ 6) Messages sent over the Internet are called _____.

_____ 7) When posting something on an Internet site, never give out your home _____ or phone number.

_____ 8) If you are home alone, keep the _____ locked and do not open it.

_____ 9) If you get a scrape while riding your bike, you probably could call your _____ for help.

_____ 10) Choking, poisoning, and drowning are the main causes of death of children under age _____.

Discover Health

Name _____ Date _____ Period _____ | Chapter 14
Workbook Activity 53

Emergency Equipment

Directions Complete each sentence by writing the letter of the best word or words from the Word Bank in the space on the left-hand side of the page.

Word Bank
a) batteries
b) candles
c) emergency
d) emergency kit
e) first aid kit
f) gas
g) radio
h) respond
i) simple
j) two

_____ 1) A collection of things you might need in an emergency is an _____.

_____ 2) A battery-powered _____ can give you information about what is happening during an emergency.

_____ 3) Radio reports tell people how to _____ in an emergency.

_____ 4) Always have at least _____ flashlights available.

_____ 5) A good source of light are _____.

_____ 6) If you smell _____ or think there's a leak, never light a match.

_____ 7) Keep new _____ near your flashlights.

_____ 8) A _____ is necessary during an emergency.

_____ 9) Bandages, scissors, tweezers, and medicines can treat _____ injuries.

_____ 10) After every _____ check your first aid kit.

Discover Health

Name _____ Date _____ Period _____

Chapter 14
Workbook Activity
54

Safety During Natural Disasters

Directions Write *T* if the statement is true or *F* if it is false.

_____ 1) An earthquake is a shaking of rocks in the earth's crust.

_____ 2) A natural gas line exploding is an example of a natural disaster.

_____ 3) The largest earthquake in U.S. history occurred in California.

_____ 4) If you are inside during an earthquake, take cover behind the curtains.

_____ 5) If you are outside during an earthquake, avoid objects that could fall on you.

_____ 6) Thunderstorms in the United States stay mainly on the Gulf Coast.

_____ 7) A hurricane is a tropical storm that forms over the ocean.

_____ 8) A flood is safe to cross if the water reaches above your knees.

_____ 9) During a tornado, do not go into the basement, and do avoid all windows.

_____ 10) A tornado usually forms over land.

_____ 11) If you are in a car during a tornado, try to stay ahead of the storm.

_____ 12) A lightning bolt can discharge a million volts of electricity.

_____ 13) The safest place to be in a thunderstorm is under a water tower.

_____ 14) If you live in a place with earthquakes, be sure heavy objects are bolted down.

_____ 15) Driving through floodwater is the safest way to cross it.

Discover Health

Name _____ Date _____ Period _____ | Chapter 15
Workbook Activity
55

What to Do First

Directions Write *T* if the statement is true or *F* if it is false.

_____ 1) First aid is emergency care given to an injured person before medical help arrives.

_____ 2) Remaining calm in an emergency is the best thing to do.

_____ 3) Good Samaritan Laws protect people who provide first aid in emergencies.

_____ 4) Rescuers are expected to risk their own lives in providing first aid.

_____ 5) Universal Precautions include wearing rubber gloves and masks.

_____ 6) Bandages that have come into contact with body fluids should be disposed of.

_____ 7) Calling 911 will put you in touch with a hospital.

_____ 8) If you or a victim are in danger, move out of the way.

_____ 9) To prevent shock, cover an accident victim with a blanket or a coat.

_____ 10) Apply direct pressure to bleeding areas to prevent further bleeding.

Discover Health

Name _____ Date _____ Period _____ | Chapter 15
Workbook Activity
56

Caring for Common Injuries

Directions Complete each sentence by writing the letter of the *best* word or words from the Word Bank in the space on the left-hand side of the page.

Word Bank

a) animal bites	f) heat exhaustion	k) rabies
b) elevate	g) hypothermia	l) snakebite
c) first	h) ligaments	m) splint
d) fracture	i) lukewarm	n) third
e) frostbite	j) nosebleed	o) three

_____ 1) A sprain is the sudden tearing of tendons or _____.

_____ 2) A cracked or broken bone is also called a _____.

_____ 3) A _____ keeps a broken limb from moving.

_____ 4) There are _____ types of burns.

_____ 5) The worst kind of burn is a _____-degree burn.

_____ 6) The best way to treat a _____ is to sit down and lean forward.

_____ 7) The opposite of heatstroke is _____ .

_____ 8) _____ results from physical exercise in very hot temperatures.

_____ 9) A disease transmitted through animal bites is _____ .

_____ 10) To treat _____ , warm the affected area of skin gradually.

_____ 11) _____ from one of the four poisonous U.S. snakes are serious.

_____ 12) ICE means immobilize, cold, and _____ .

_____ 13) If someone's eye has an object in it, flush the eye with _____ water.

_____ 14) A sunburn is an example of a _____-degree burn.

_____ 15) For all _____ , wash the area with soap and warm water and apply a dressing.

Discover Health

Name _____ Date _____ Period _____

Chapter 15
Workbook Activity
57

First Aid for Bleeding, Shock, and Choking

Directions Complete each sentence by writing the letter of the *best* word or words from the Word Bank in the space on the left-hand side of the page.

Word Bank

a) decrease
b) four
c) finger sweep
d) Heimlich maneuver
e) patting
f) physical
g) pressure
h) rubber gloves
i) shock
j) spine

_____ 1) A rescuer needs to do _____ things to control severe bleeding.

_____ 2) To _____ blood flow to a wound, elevate the wound above the heart.

_____ 3) Slow pulse, slow or fast breathing, pale skin, thirst, or confusion are signs of _____ .

_____ 4) Upward abdominal thrusts are part of the _____ .

_____ 5) _____ the back of a person who is choking may make the object go farther down the throat.

_____ 6) Use a _____ to check if something is blocking a choking victim's throat.

_____ 7) In treating severe bleeding, always protect yourself with _____ , if they are available.

_____ 8) Shock is a _____ reaction to injury.

_____ 9) If you think an injured person may have a neck or _____ injury, do not move the person.

_____ 10) To help stop severe bleeding, apply direct _____ with the palm of your hand.

Discover Health

Name _____ Date _____ Period _____

Chapter 15
Workbook Activity
58

First Aid for Heart Attacks and Poisoning

Directions Write *T* if the statement is true or *F* if it is false.

_____ 1) Cardiotachography is an emergency procedure to treat cardiac arrest.

_____ 2) Cardiac arrest is a condition in which the heart stops beating.

_____ 3) Oral poisoning occurs when someone touches a poison.

_____ 4) Contact poisoning occurs when someone touches an electrical circuit.

_____ 5) To treat contact poisoning, wash the affected area.

_____ 6) To treat inhalation poisoning, induce vomiting.

_____ 7) If someone swallows poison, call the poison control center at once.

_____ 8) If a person is in cardiac arrest, the person has no pulse.

_____ 9) Pressing the chest during CPR helps pump blood to an injured person's body.

_____ 10) Poisons can be absorbed through the skin from certain plants, cleaning products, or lawn chemicals.

Discover Health

Name _____ Date _____ Period _____

Chapter 16
Workbook Activity
59

Defining Violence

Directions Write *T* if the statement is true or *F* if it is false.

_____ 1) The media are sources of entertainment, such as newspapers or TV.

_____ 2) Problems are actions or words that hurt people or things they care about.

_____ 3) A disagreement or difference of opinion is conflict.

_____ 4) A cycle is having two ideas and not being sure which is best.

_____ 5) A repetition is also called a cycle.

_____ 6) Everyone pays for the costs of violence.

_____ 7) Violence that happens on TV, video games, or movies is media violence.

_____ 8) Random violence happens in a family.

_____ 9) Aggression is violence that is not directed at a specific person.

_____ 10) Violence is an effective way to solve problems.

Discover Health

Name _____ Date _____ Period _____

Chapter 16
Workbook Activity
60

Causes of Violence

Directions Put a check mark next to the *best* answer.

Column A

_____ 1) warning signs of conflict

_____ 2) nearly half of all crimes are committed in this age group

_____ 3) opinion based on someone's race, religion, or culture

_____ 4) tip for staying away from dangerous strangers

_____ 5) one reason people act violently

_____ 6) parents' violence sets this for their children

_____ 7) having a gun in the home increases this chance

_____ 8) where child abuse often happens

_____ 9) one cause of people's dangerous behavior

_____ 10) taking revenge may lead to this

Column B

a) 12 to 24

b) accidental death

c) drugs and alcohol

d) poor example

e) prejudice

f) shouting, insults, name-calling

g) to show loyalty to a group

h) violence

i) walk only in familiar areas

j) within a family

Discover Health

Name _____ Date _____ Period _____

> Chapter 16
> Workbook Activity
> **61**

Preventing Violence

Directions Write *T* if the statement is true or *F* if it is false.

_____ 1) A mediator is someone who helps two sides solve a problem.

_____ 2) By favoring one side over another, you can remain neutral.

_____ 3) One way to relax is to remain calm and focus on your own feelings.

_____ 4) If someone is strong enough to apologize, help them by giving them respect.

_____ 5) You may need a parent or other adult to help solve a serious conflict.

_____ 6) Keeping a secret is more important than stopping violence.

_____ 7) If you know a person may hurt others, tell a responsible adult.

_____ 8) When resolving a conflict, try to see it from the other person's viewpoint.

_____ 9) In solving a conflict, there are methods you can use.

_____ 10) When listening to someone explain a problem, do not interrupt or act bored.

Discover Health

Name _____ Date _____ Period _____

Chapter 17
Workbook Activity
62

Health Care Information

Directions Write *T* if the statement is true or *F* if it is false.

_____ 1) Health care includes medical procedures and emergency treatments.

_____ 2) You need a doctor's written order to buy prescription medicine.

_____ 3) A general practitioner is an example of a specialist.

_____ 4) Health care includes having a healthy diet.

_____ 5) Managed health care may require you to see a doctor in the company's plan.

_____ 6) Only a few health care products can help diagnose and treat illnesses.

_____ 7) Health care includes preventive medicine.

_____ 8) Employees at a health care insurance company are part of the health care industry.

_____ 9) Workers at companies that make health care products are part of the health care industry.

_____ 10) Health care includes oversleeping twice a week.

Discover Health

Name _____ Date _____ Period _____ | Chapter 17
Workbook Activity
63

Seeking Health Care

Directions Write *T* if the statement is true or *F* if it is false.

_____ 1) Prevention is appropriate, ongoing self-care.

_____ 2) A nurse practitioner is a registered nurse with special training for providing health care.

_____ 3) A doctor who treats people for routine problems is called a specialist.

_____ 4) Rehabilitation is help to recover from surgery, illness, or injury.

_____ 5) Pediatrics is child health care.

_____ 6) Inpatient care is health care received without staying overnight at a health care facility.

_____ 7) Inpatient care is health care received while staying overnight at a care facility.

_____ 8) A health care facility is a place people can go for medical, dental, or other care.

_____ 9) A long-term care facility for people who are dying is called a hospital.

_____ 10) Someone who is dying from disease, injury, or illness, sometimes over a long period, is said to be terminally ill.

Discover Health

Name _____ Date _____ Period _____ | Chapter 17
Workbook Activity
64

Paying for Health Care

Directions Write *T* if the statement is true or *F* if it is false.

_____ 1) The initial amount you pay before your health insurance covers health care costs is a conductant.

_____ 2) Outpatient care is almost always better than inpatient care.

_____ 3) A government health care plan is called Medicare.

_____ 4) Medicare is for people age 65 and older who receive Social Security.

_____ 5) Medicaid provides health care for people with incomes below a certain amount.

_____ 6) You pay a premium to receive health care.

_____ 7) Health insurance is meant to cause out-of-pocket expenses to rise.

_____ 8) Preventive medicine is the most expensive form of health care.

_____ 9) Outpatient care requires a patient to stay overnight at a hospital.

_____ 10) Health insurance is a system where you pay less money the longer you stay in a hospital.

Discover Health

Name _____ Date _____ Period _____

Chapter 17
Workbook Activity 65

Being a Wise Consumer

Directions Write the letter of the item from Column B that *best* matches the description in Column A.

Column A

_____ 1) someone who buys goods or services

_____ 2) the number of items in a package

_____ 3) not working properly

_____ 4) a medical product or service that is unproved or worthless

_____ 5) nongenuine

_____ 6) information about a product that you hear from someone you know

_____ 7) best source of medical information

_____ 8) what a wise consumer has

_____ 9) what advertising does to a consumer's choices

_____ 10) nonbrand-name

Column B

a) bogus

b) consumer

c) count

d) defective

e) doctor or pharmacist

f) generic

g) increased self-confidence

h) influences

i) quackery

j) word of mouth

Discover Health

Chapter 17
Workbook Activity
66

Evaluating Advertisements

Directions Write *T* if the statement is true or *F* if it is false.

_____ 1) Companies rarely spend large amounts of money to advertise products.

_____ 2) Sometimes products advertised as new really aren't.

_____ 3) People may buy a product that reminds them of their past.

_____ 4) The more expensive a product is, the better it must be.

_____ 5) Consumers can always depend on advertising to give the best information about a product.

_____ 6) A cartoon character might influence children to buy a product.

_____ 7) Famous people are rarely used to sell products.

_____ 8) Someone over age 21 is called a minor.

_____ 9) Companies make claims about their products to sell more than the competition.

_____ 10) The U.S. government has removed all restrictions on advertising cigarettes.

Discover Health

Name _____ Date _____ Period _____

Chapter 17
Workbook Activity
67

Consumer Protection

Directions Write *T* if the statement is true or *F* if it is false.

_____ 1) The U.S. government established the Patient's Bill of Rights.

_____ 2) The U.S. government established the Consumer Bill of Rights.

_____ 3) Consumer advocates represent companies that advertise their products.

_____ 4) The first step to correcting a consumer problem is to contact the Better Business Bureau.

_____ 5) As a patient, you have the right to considerate and respectful care.

_____ 6) As a consumer, you have the right to be informed.

_____ 7) If contacting a manufacturer or caregiver doesn't solve your problem, then you can contact a government agency like the Food and Drug Administration.

_____ 8) The government can help you take legal action against a manufacturer.

_____ 9) The American Hospital Association established the Patient's Bill of Rights.

_____ 10) If a manufacturer doesn't respond to your request or problem in two days, send a follow-up letter.

Discover Health

Name _____ Date _____ Period _____

Chapter 18
Workbook Activity
68

Defining Community

Part A Directions Place a check mark next to the *best* answer.

1) Which of these are communities?
_____ **a)** a small village
_____ **b)** a neighborhood in a large city
_____ **c)** an ethnic or religious group
_____ **d)** all of the above

2) Communities are formed by people with a common:
_____ **a)** location.
_____ **b)** identity.
_____ **c)** association.
_____ **d)** all of the above

3) Health is a community concern because people:
_____ **a)** with communicable diseases can infect others.
_____ **b)** need to know about and use available health services.
_____ **c)** without health care may need assistance.
_____ **d)** all of the above

4) What does the World Health Organization (WHO) do?
_____ **a)** provides assistance to combat diseases that spread globally
_____ **b)** monitors world health problems
_____ **c)** promotes research and training in the area of health
_____ **d)** all of the above

Part B Directions Write the letter of the item from Column B that *best* matches the description in Column A.

Column A	Column B
_____ **5)** One way a community may show pride	**a)** communicable
_____ **6)** How money and resources are divided	**b)** economics
_____ **7)** Source of supply and support	**c)** communicable
_____ **8)** Able to be passed to others	**d)** holding a festival
_____ **9)** Disease that spreads quickly	**e)** poverty
_____ **10)** Major health concern because it puts people at risk for health problems	**f)** resource

Discover Health

Name _____ Date _____ Period _____ | Chapter 18
Workbook Activity 69

Community Health Resources

Part A Directions Write the letter of the item from Column B that matches the description in Column A.

Column A

_____ 1) private charity that gives money for health programs

_____ 2) consists of people who give their time to organizations such as the American Cancer Society, and groups of people with similar health problems or concerns, such as the Center for Independent Living

_____ 3) a large federal department that is concerned with the health and well-being of the nation

_____ 4) an agency responsible for conducting research concerning the nation's health problems by providing grants for research

_____ 5) tests the foods, drugs, and cosmetics we use and requires food manufacturers to label ingredients on food containers

_____ 6) the agency that works to prevent the spread of disease in the United States by tracking the spread of infectious diseases

_____ 7) the agency that deals with the safety of products by investigating any that are thought to be defective

_____ 8) important information

Column B

a) Centers for Disease Control (CDC)

b) charitable organization

c) Consumer Product Safety Commission

d) Department of Health and Human Services (DHHS)

e) Food and Drug Administration (FDA)

f) National Institutes of Health (NIH)

g) statistics

h) volunteer resources

Part B Directions Place a check mark next to the *best* answer.

9) Which of the following is a private resource that provides health services?
_____ a) local health department
_____ b) health insurance company
_____ c) American National Red Cross
_____ d) none of the above

10) Public health resources are a part of the _____ government.
_____ a) federal
_____ b) state
_____ c) local
_____ d) all of the above

Discover Health

Community Health Advocacy Skills

Directions Write *T* if the statement is true or *F* if it is false.

_____ 1) Taxes pay for community services.

_____ 2) Community services can tell you about agencies that can help in emergencies.

_____ 3) A budget is a written plan for managing money.

_____ 4) A budget will not help you plan how to spend money.

_____ 5) Always try not to spend more than your income.

_____ 6) Fixed expenses are more important than others are.

_____ 7) Renting a video game is an example of a fixed expense.

_____ 8) It is better to treat an illness than prevent it.

_____ 9) There are many ways to contribute to your community.

_____ 10) An advocacy group tries to help anyone in need.

Discover Health

Name _____ Date _____ Period _____ | Chapter 19
Workbook Activity
71

Health and the Environment

Directions Write *T* if the statement is true or *F* if it is false.

_____ 1) The system that connects air, land, water, plants, and animals is called ecology.

_____ 2) A buildup of harmful wastes in air, land, or water is pollution.

_____ 3) The environment is the study of how living things are connected.

_____ 4) The natural balance may be disturbed by natural events or by human activity.

_____ 5) The only animals who have the ability to change the environment are humans.

_____ 6) As more people inhabit the earth, more demands are placed on humans.

_____ 7) The earth's resources are unlimited.

_____ 8) Volcanic eruptions, floods, and earthquakes are examples of natural disasters.

_____ 9) Metals, minerals, coal, and oil are examples of resources that cannot be replaced.

_____ 10) In response to natural disasters, the environment will change.

Discover Health

Name _____ Date _____ Period _____ | Chapter 19
Workbook Activity
72

Air Pollution and Health

Directions Complete each sentence by writing the letter of the *best* word or words from the Word Bank in the space on the left-hand side of the page.

Word Bank
a) acid rain   **f)** chlorofluorocarbons (CFCs)   **k)** ozone layer
b) asbestos   **g)** Environmental Protection Agency (EPA)   **l)** particulate
c) by-product   **h)** fossil fuel   **m)** radon
d) cancer   **i)** greenhouse effect   **n)** smog
e) carbon monoxide   **j)** hydrocarbon   **o)** toxin

_____ 1) rain, snow, sleet, or hail with large amounts of sulfuric acid

_____ 2) tiny pieces of solid matter such as dust, ash, or dirt in the air

_____ 3) a material made of fibers, that was once used in insulating materials and that can cause cancer

_____ 4) a pollutant caused by car exhaust

_____ 5) government agency that can help you remove radon from your home

_____ 6) an unwanted result

_____ 7) a toxin caused by motor vehicle exhaust

_____ 8) a condition in which a carbon dioxide cloud acts like a glass ceiling, trapping heat and moisture and resulting in the warming of the earth's atmosphere

_____ 9) burnable substance formed from plant or animal material

_____ 10) a colorless, odorless, poisonous gas that is formed underground

_____ 11) a poisonous chemical

_____ 12) region in the atmosphere that protects the earth from the sun's rays

_____ 13) may be responsible for reduction in ozone layer

_____ 14) air pollution formed by car exhaust and other pollutants

_____ 15) disease that may be caused by asbestos fibers

Discover Health

Name _____ Date _____ Period _____

> Chapter 19
> Workbook Activity
> **73**

Water and Land Pollution and Health

Directions Write *T* if the statement is true or *F* if it is false.

_____ 1) People need freshwater so that they can go boating.

_____ 2) If groundwater is contaminated, then toxins may be transferred to people.

_____ 3) Toxic by-products pollute the water supply and hurt marine life.

_____ 4) To treat raw sewage, most cities ship it to Europe.

_____ 5) Some household chemicals that are used every day can harm the water supply.

_____ 6) Two threats to groundwater are overuse of water towers and rundown reservoirs.

_____ 7) One way to conserve our groundwater is to let the water run while you brush your teeth.

_____ 8) Famine is an illness in a family.

_____ 9) A place where waste is buried between layers of earth is called a landfill.

_____ 10) To keep the earth healthy, you should buy products from South America.

Discover Health

Name _____ Date _____ Period _____

Chapter 19
Workbook Activity
74

Promoting a Healthy Environment

Directions Place a check mark beside the *best* answer.

1) The _____ has passed laws that protect the environment.
   _____ a) U.S. government   _____ b) president of the United States
   _____ c) PTA   _____ d) National Pollution Center

2) The Clean Air Act sets limits on the levels of _____
   _____ a) pollutants.   _____ b) automobiles.
   _____ c) noise pollution.   _____ d) leaf burning.

3) When the air is _____ , cities sometimes issue smog alerts.
   _____ a) clean   _____ b) dark
   _____ c) dirty   _____ d) breathable

4) Scientists are looking for alternative sources of energy that _____ the earth.
   _____ a) pollute   _____ b) use resources of
   _____ c) do not pollute   _____ d) exhaust

5) Electric cars pollute _____ regular cars.
   _____ a) more than   _____ b) as much as
   _____ c) less than   _____ d) just like

6) By growing more food in smaller areas, _____ may be changed back to wetlands.
   _____ a) downtown areas   _____ b) landfills
   _____ c) farmland   _____ d) deserts

7) Individuals can have a great effect on improving _____
   _____ a) environment.   _____ b) average score.
   _____ c) stock market index.   _____ d) work efficiency.

8) Solar energy is an example of an _____
   _____ a) way of life.   _____ b) energy source.
   _____ c) landfill.   _____ d) transportation method.

9) Recycling is a good way to help keep the _____ clean.
   _____ a) environment   _____ b) bathroom
   _____ c) bathtub   _____ d) park

10) A mixture of _____ can be used to clean your windows.
    _____ a) oil and water   _____ b) vinegar and oil
    _____ c) vinegar and water   _____ d) soap and oil

Discover Health